Capybaras

By Alexandria Manera

Steadwell Books

Raintree
A Division of Reed Elsevier, Inc.

Chicago, Illinois
www.raintreelibrary.com

ANIMALS OF THE RAIN FOREST

©2003 Raintree
Published by Raintree, a division of Reed Elsevier Inc., Chicago, Illinois

For information, address the publisher:
Raintree, 100 N. LaSalle, Suite 1200, Chicago, IL 60602

Library of Congress Cataloging-in-Publication Data
Manera, Alexandria.
   Capybaras / Alexandria Manera.
      v. cm. -- (Animals of the rain forest)
Includes bibliographical references (p.   ).
Contents: Range map for capybaras -- Quick look at capybaras -- Capybaras in the rain forest -- What capybaras eat -- A capybara's life cycle.
>>   ISBN 0-7398-6835-7 (lib. bdg.-hardcover)
  1. Capybara--Juvenile literature. [1. Capybara.] I. Title. II.Series.
   QL737.R662 M36 2003
599.35'9--dc21

                                   2002015208
Printed and bound in the United States of America

Produced by Compass Books

Photo Acknowledgments

Tom Stack/Joe McDonald, cover, 1; Luiz Claudio Margio, 6, 8, 12, 21; Root Resources/Kenneth W. Fink, 11; Unicorn/Tommy Dodson, 14; Visuals Unlimited/ Inga Spence, 16; Visuals Unlimited/Albert Copely, 18; Root Resources/Mary and Lloyd McCarthy, 22; Wildlife Conservation Society/Elyssa Kellerman 24; Wildlife Conservation Society, D. Demello, 26.

Content Consultants

Dr. Emilio Herrera
Department of Environmental Studies
Universidad Simon Bolivar
Caracas, Venezuela

Mark Rosenthal
Abra Prentice Wilkin Curator of Large Mammals
Lincoln Park Zoo
Chicago, Illinois

This book supports the National Science Standards.

Some words are shown in bold, **like this**. You can find out what they mean by looking in the Glossary.

Contents

Range Map of Capybaras. 4

A Quick Look at Capybaras. 5

Capybaras in the Rain Forest. 7

What Do Capybaras Eat. 15

A Capybara's Life Cycle. 19

The Future of Capybaras. 25

Photo Diagram. 28

Glossary. 30

Internet Sites, Address, Books to Read. 31

Index. 32

Range Map of Capybaras

MEXICO

BELIZE
HONDURAS

GUATEMALA
EL SALVADOR

NICARAGUA

Caribbean
Sea

North
Atlantic
Ocean

GUYANA
SURINAME

COSTA RICA

PANAMA

VENEZUELA

FRENCH
GUIANA
(FRANCE)

ECUADOR

COLOMBIA

PERU

AMAZON
RIVER

BRAZIL

BOLIVIA

South
Pacific
Ocean

CHILE

PARAGUAY

South
Atlantic
Ocean

ARGENTINA

URUGUAY

Range of the
Capybaras

Surrounding
Land

Water

Borders

Rivers

N
W   E
S

A Quick Look at Capybaras

What do capybaras look like?

Capybaras have large heads and snouts. Their small eyes and ears are on the tops of their heads. Very little hair covers their thick bodies. They have four short legs and webbed feet.

Where do capybaras live?

Capybaras live in the rain forests and in wet, grassy areas of South America. They make their homes near lakes, rivers, and swamps.

What do capybaras eat?

Capybaras eat grasses and plants that grow in or near the water. Sometimes they eat grains, melons, and squashes.

Capybaras are excellent swimmers and spend a lot of time in the water.

Capybaras in the Rain Forest

Capybaras (kap-uh-bar-UHZ) are the largest rodents in the world. All rodents have sharp front teeth called incisors. These teeth never stop growing. Mice and rats are other kinds of rodents.

Rodents are mammals. Mammals have fur and supply milk to their young. They are warm-blooded, which means their body temperatures stay the same in hot or cold surroundings. Like other mammals, rodents also breathe air and give live birth.

Scientists call capybaras *Hydrochaerus hydrochaeris*. This scientific name means water pig. Capybaras spend a lot of time in the water. Even though their name means water pig, they are actually rodents.

Capybaras live near watery places in tropical grasslands and rain forests.

Where do capybaras live?

Capybaras live in Central and South America. They are common from Panama down to Argentina. Many capybaras live along the Pantanal in Brazil and the Llanos in Colombia and Venezuela. These are large grasslands that flood for part of the year. There are many rivers and forests in the Pantanal and the Llanos.

A capybara's **habitat** has a lot of water. A habitat is an area where a plant or animal usually lives. Capybaras most often live near lakes, streams, or swamps. These watery places can be in rain forests or open grasslands.

Capybaras spend most of their lives in one area called a home range. A capybara's home range is from 24 to 49 acres (10 to 20 hectares). A group of about 10 to 20 capybaras lives in an area. They can find food and everything they need to live in their home range.

Water is always a part of a capybara's home range because they are **semi-aquatic**. This means that it spends part of its life in the water. They are excellent swimmers.

Capybaras sometimes use the water for protection from predators. Predators are animals that hunt and eat other animals. Capybaras' predators are the jaguar, caiman, and anaconda.

What do capybaras look like?

Capybaras look like very large guinea pigs. Their bodies are thick and three to four feet (1-1.3 meters) long. They stand nearly two feet (60 centimeters) tall.

Most capybaras weigh about 100 pounds (45 kilograms) but can sometimes weigh as much as 175 pounds (80 kilograms). One male often leads a group of capybaras. He is usually larger than the others in the group.

Capybaras have wide heads and large snouts. Their eyes, noses, and ears are set high on their heads. This helps them to see, breathe, and hear while in the water.

Usually, capybaras are brown or red in color. Their bellies are sometimes yellow. Their hair is long and spread thinly over their skin. Most of their skin is bare. Capybaras take a swim or find shade from a nearby tree to avoid getting burned by the sun.

Capybaras have four webbed toes on their front feet and three webbed toes on their hind feet. Webbed toes have a flap of skin that connects each toe. This helps them to swim quickly and to walk on the soggy ground.

> **You can see the skin growing between this capybara's toes. Webbed feet help it swim.**

Their front legs are slightly shorter than their hind legs. This helps them to dig while on land.

Sometimes capybaras dig **wallows**. A wallow is a shallow ditch. Capybaras usually dig wallows in soft, wet soil. Water and mud collect on the bottom of the wallow. Then, capybaras roll around in the wallow. They may also rest there. The mud and water helps to keep them cool and protects their skin from the sun.

> **This herd of capybaras is drinking and swimming in a water hole.**

How do capybaras act?

Capybaras mainly live in small family groups of ten to twenty. One dominant male is the leader of the group. This group will also have several younger males, females, and babies.

During the dry season, when there are fewer watery places, a group might gather with other small groups to form a herd. Large numbers of capybaras gather around each pool of water. There can be up to 100 capybaras in a herd, although this large number is not common.

These large herds stay together for protection. They will bark to alarm the group when danger is near. They will often dive into the water to hide from their predators. Capybaras can stay underwater for almost five minutes.

Capybaras are social and peaceful animals. Social animals enjoy being in groups. Capybaras do not often fight. Sometimes capybara groups will defend their territory from neighboring groups and predators. They do this most often during the wet season.

Male capybaras have a scent gland called a Morrillo gland on their snout. A gland is a part of the body that releases a useful substance. The Morrillo gland releases a white paste. The male capybara will rub it onto females or nearby plants. This is how some males mark their territory.

Capybaras can graze on land or in the water when grasslands are flooded.

What Do Capybaras Eat?

Capybaras are **herbivores**. Herbivores eat only plants. Capybaras usually eat grasses and other plants that grow in the water. The name "capybaras" means master of the grasses. Sometimes capybaras will eat crops such as grains, melons, or squashes.

Capybaras choose their food carefully. They will eat only certain types of plants. Adult capybaras eat nearly six to eight pounds (three to four kilograms) of grass a day.

Capybaras will not graze in the same place two days in a row. Graze means to eat grass. They search their range for grasses that have not been touched. This gives the other plants time to grow again. This is good for their environment, or natural surroundings.

Capybaras live in areas with plenty of plants. This makes it easy for them to find food.

Digestion

Capybaras use their incisors to **gnaw**, or bite, their food into fine pieces. Like other **rodents**, capybaras have four incisors. These teeth are always growing. They will continue to grow throughout the capybara's life. Capybaras must chew tough grasses to keep their teeth from becoming too long.

It is hard for mammals to **digest** grass. Digestion is the process where animals break down food so they can absorb the nutrients they need. Capybaras need special bacteria to help them get nutrients from the plants they eat. To help them do this, capybaras practice coprophagy. Coprophagy means that capybaras eat their droppings. The capybara digests nutrients as the droppings pass through its digestive system for the second time.

Capybaras mate in the water.

A Capybara's Life Cycle

Capybaras can mate anytime throughout the year but usually mate before the rainy season begins. In Colombia and Venezuela, mating is most common in April or May. In Brazil, they usually mate around November.

A female capybara's scent changes when she is ready to mate. She also lets out whistling noises through her nose.

Capybaras mate in the water. The female often chooses her mate. Males court females by following them around. They go in and out of the water several times. If a female does not want to mate,.she will leave the water.

The dominant male usually mates with the females in his group. He may stop other males from mating. He does this by swimming in between the male and the female.

Young

It takes female capybaras almost 5 months (150 days) until their litter is born. Female capybaras usually have 4 or 5 babies at one time. The mother will return to the group a few hours after the young are born.

Baby capybaras are well developed when they are born. They can see and walk after birth. Within a day or two, mother and babies will return to the group. Baby capybaras are able to eat and **digest** grass after one week.

Capybara babies group together in crèches. A crèche is a group of young. Once in a crèche, it is hard for the mother to tell which are her young. Sometimes, a female will nurse other female's babies. To nurse is to feed milk to the young.

Like the adults, young capybaras are very vocal. This means they make many sounds. They begin to communicate by purring. They learn to communicate from the others in the group.

Capybaras stay with their mothers for almost one year. They stay in the same group, but leave their mothers to mate. Wild capybaras usually live for eight to ten years.

This mother capybara is taking care of her young babies.

> These capybaras are cooling off in a wallow that they have dug.

A capybara's day

Capybaras are **crepuscular**, which means they are active during twilight hours. Twilight is the time between sunset and dark. It is often too hot for capybaras to be active in the daytime. They search for food when it is cool.

Capybaras graze on the **shoal** in the early morning. A shoal is a stretch of shallow water. They also rest in **wallows**.

In the afternoon, capybaras will swim or rest near the water. They often use the water to cool off during the hot afternoon. The water helps protect their skin from the powerful sun.

At night, capybaras go back to grazing or resting on the shoal. Capybaras are even nocturnal in areas where they are hunted. Nocturnal means they are active at night. Some people hunt them for their meat and skin.

Capybaras take care of their young. The mother and the other females watch the young closely. Adults stop grazing or feeding their young if a predator is near. They form a tight circle around their young to protect them. They may also run away or dive into the water to hide.

Scientists do not know exactly how many capybaras are left in the wild.

The Future of Capybaras

Capybaras are not an **endangered species**. Endangered means that an animal is in danger of dying out. However, if people continue to destroy the rain forest they may become endangered.

Some people are destroying the rain forests and grasslands. They even drain, or empty, the lakes, ponds, and rivers where capybaras live. They use the land for growing crops, raising cattle or mining. Capybaras lose their natural **habitat** when people ruin the land.

Humans also hunt capybaras for their meat and skin. The number of capybaras has become smaller because of overhunting. Overhunting means that people hunt too many animals.

These capybaras are eating food that people have placed outside for them.

What will happen to capybaras?

Some people are trying to preserve the rain forests and the animals in them. Many national parks and reserves have been created. Reserves are areas of land that are protected. People make laws to protect the land and stop the destruction of the rain forest and grasslands.

Some people are building ranches to protect capybaras. Ranches are large areas of land, often near national parks. They help to stop people from over-hunting capybaras. They also help to preserve the capybara's natural **habitat**.

Capybaras are an important part of the rain forest and grasslands. Some people believe that their careful grazing helps to preserve their habitat. These people are trying to protect capybaras in the wild.

There are many capybaras living throughout South America. People must work together to preserve the rain forest and tropical grassland habitats. That way, there will be many capybaras in the future.

hair
see page 10

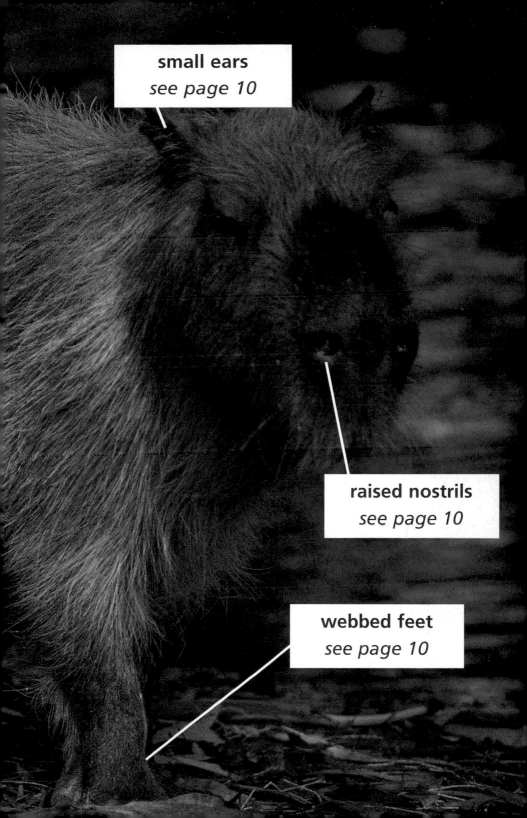

small ears
see page 10

raised nostrils
see page 10

webbed feet
see page 10

Glossary

crepuscular—active during the twilight hours

digestion—the process of breaking down food in the stomach so it can be absorbed into the blood

endangered species—a type of plant or animal that is in danger of becoming extinct

gnaw—to keep biting on something

habitat—the place where a plant or animal lives

herbivore—an animal that eats plants rather than other animals

regurgitate—to bring food from the stomach back into the mouth

rodent—a mammal with large, sharp front teeth that it uses for gnawing

semi-aquatic—living partly in the water

shoal—a stretch of shallow water

wallow—a shallow ditch, often filled with mud or water, that an animal digs

Internet Sites

Amazon Conservation Team for Kids
http://www.amazonteam.org/actnew/kids.html

The Capybara Page
http://www.rebsig.com/capybara

Useful Address

Amazon Conservation Team
4211 N. Fairfax Drive
Arlington, VA 22203

Books to Read

Fitzsimons, Cecilia. *Animal Habitats*. Austin, TX: Raintree Steck-Vaughn, 1996.

Harris, Nicholas. *Into the Rain Forest: One Book Makes Hundreds of Pictures of Rain Forest Life.* Alexandria, VA: Time-Life Books, 1996.

Index

anaconda 9

Brazil 4, 8, 19

caiman 9

Colombia 4, 8, 19

coprophagy 17

crèche 20

digestion 17, 20, 30

graze 14, 15, 23

habitat 9, 25, 27, 30

herbivore 15, 30

jaguar 9

litter 20

mammals 7, 17

Morrillo gland 13

rodent 7, 17, 30

shoal 23, 30

South America 4, 5, 8, 27

Venezuela 4, 8, 19

water 5, 7, 9-15, 18, 19, 23